I wanna be a boy!

GENE MACHINES

Fran Balkwill & Mic Rolph

Cold Spring Harbor Laboratory Press

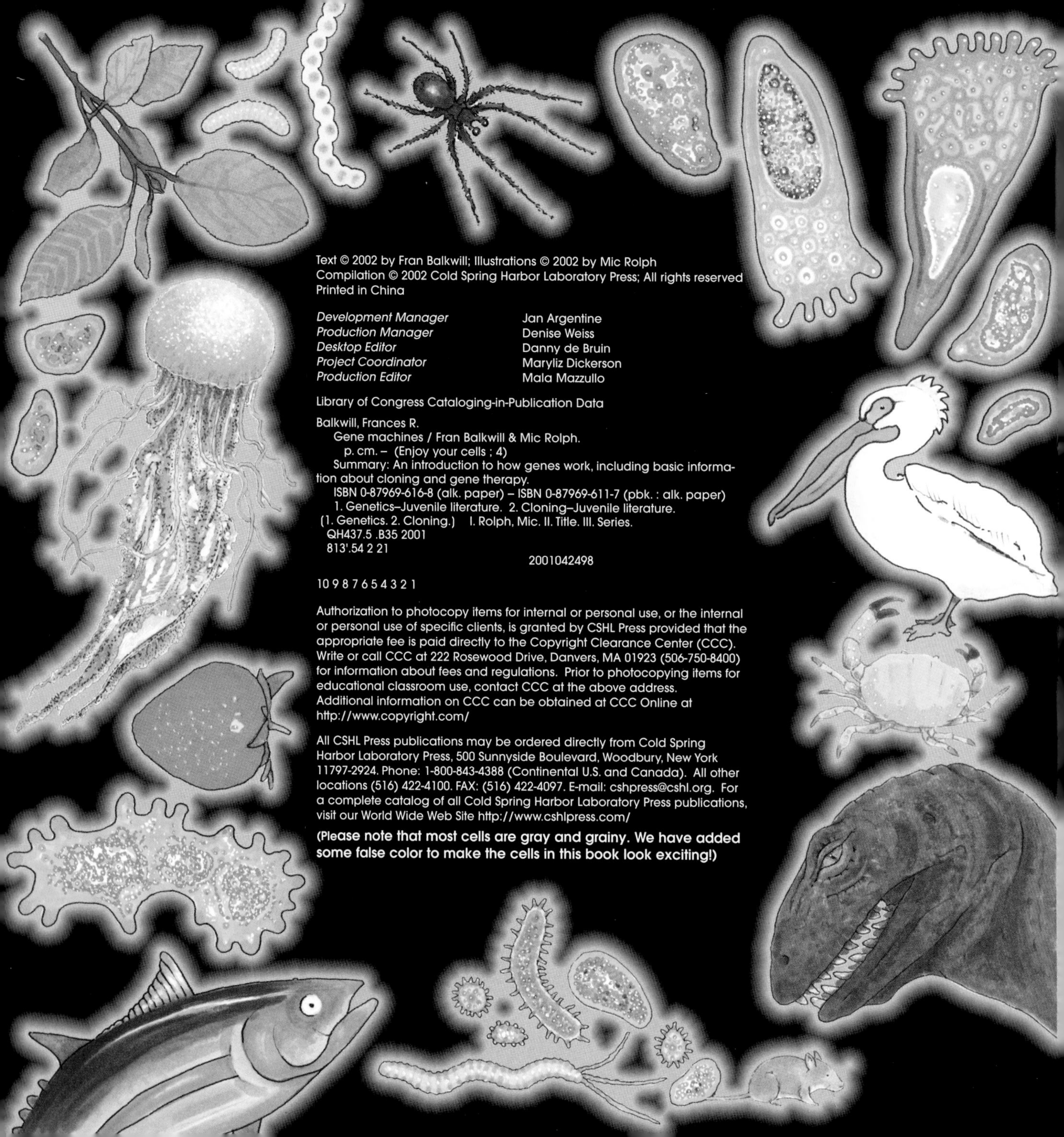

Printed in China

Development Manager Jan Argentine
Production Manager Denise Weiss
Desktop Editor Danny de Bruin
Project Coordinator Maryliz Dickerson
Production Editor Mala Mazzullo

Library of Congress Cataloging-in-Publication Data

Balkwill, Frances R.
Gene machines / Fran Balkwill & Mic Rolph.
p. cm. – (Enjoy your cells ; 4)
Summary: An introduction to how genes work, including basic information about cloning and gene therapy.
ISBN 0-87969-616-8 (alk. paper) – ISBN 0-87969-611-7 (pbk. : alk. paper)
1. Genetics–Juvenile literature. 2. Cloning–Juvenile literature.
(1. Genetics. 2. Cloning.) I. Rolph, Mic. II. Title. III. Series.
QH437.5 .B35 2001
813'.54 2 21
2001042498

10 9 8 7 6 5 4 3 2 1

All CSHL Press publications may be ordered directly from Cold Spring Harbor Laboratory Press, 500 Sunnyside Boulevard, Woodbury, New York 11797-2924. Phone: 1-800-843-4388 (Continental U.S. and Canada). All other locations (516) 422-4100. FAX: (516) 422-4097. E-mail: cshpress@cshl.org. For a complete catalog of all Cold Spring Harbor Laboratory Press publications, visit our World Wide Web Site http://www.cshlpress.com/

(Please note that most cells are gray and grainy. We have added some false color to make the cells in this book look exciting!)

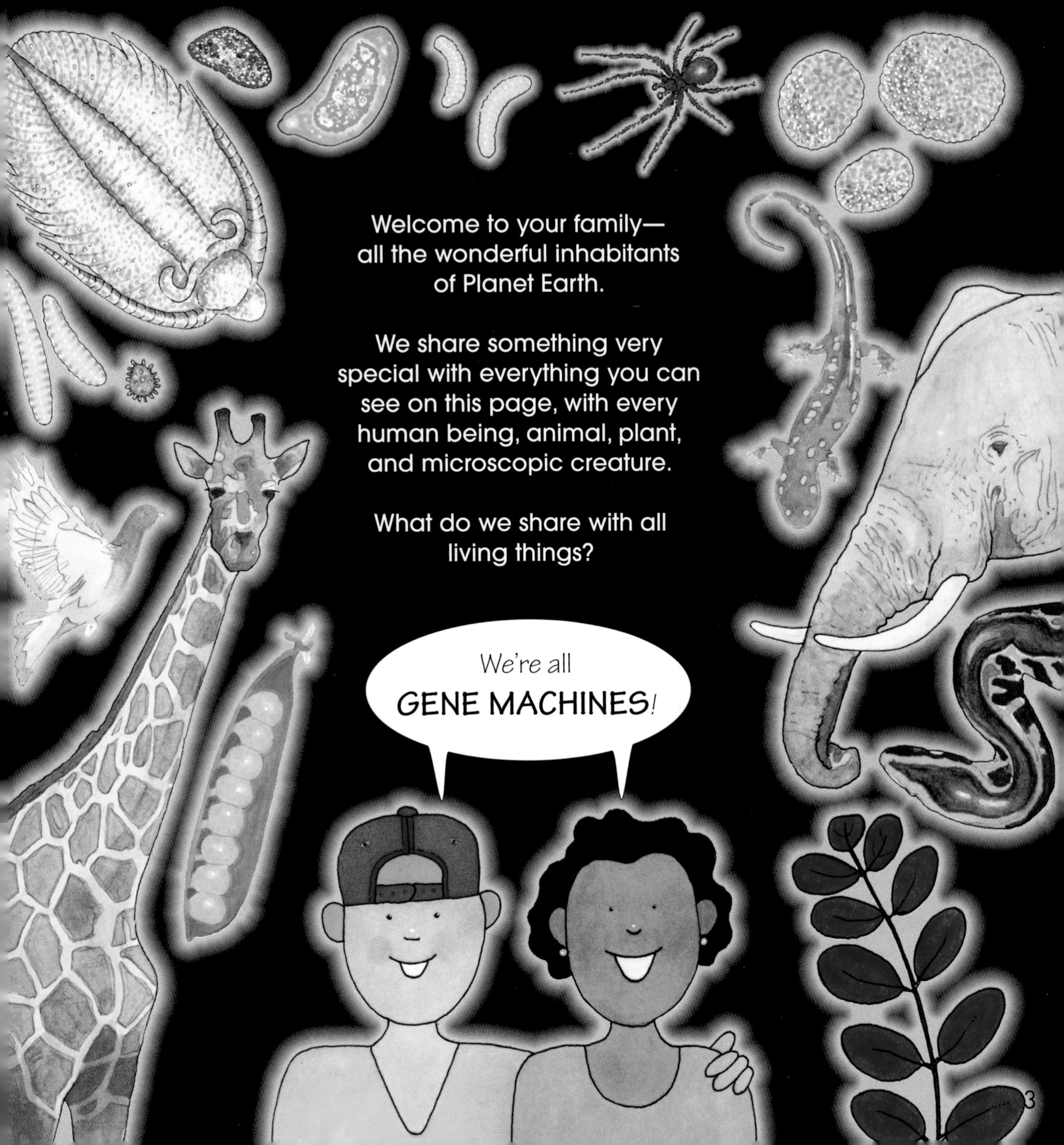
Welcome to your family—
all the wonderful inhabitants
of Planet Earth.
We share something very
special with everything you can
see on this page, with every
human being, animal, plant,
and microscopic creature.
What do we share with all
living things?
We're all
GENE MACHINES!

You are made of millions and millions of microscopic cells, each one much smaller than a grain of sand.

Inside each of those cells, there is a nucleus (new-clee-us).

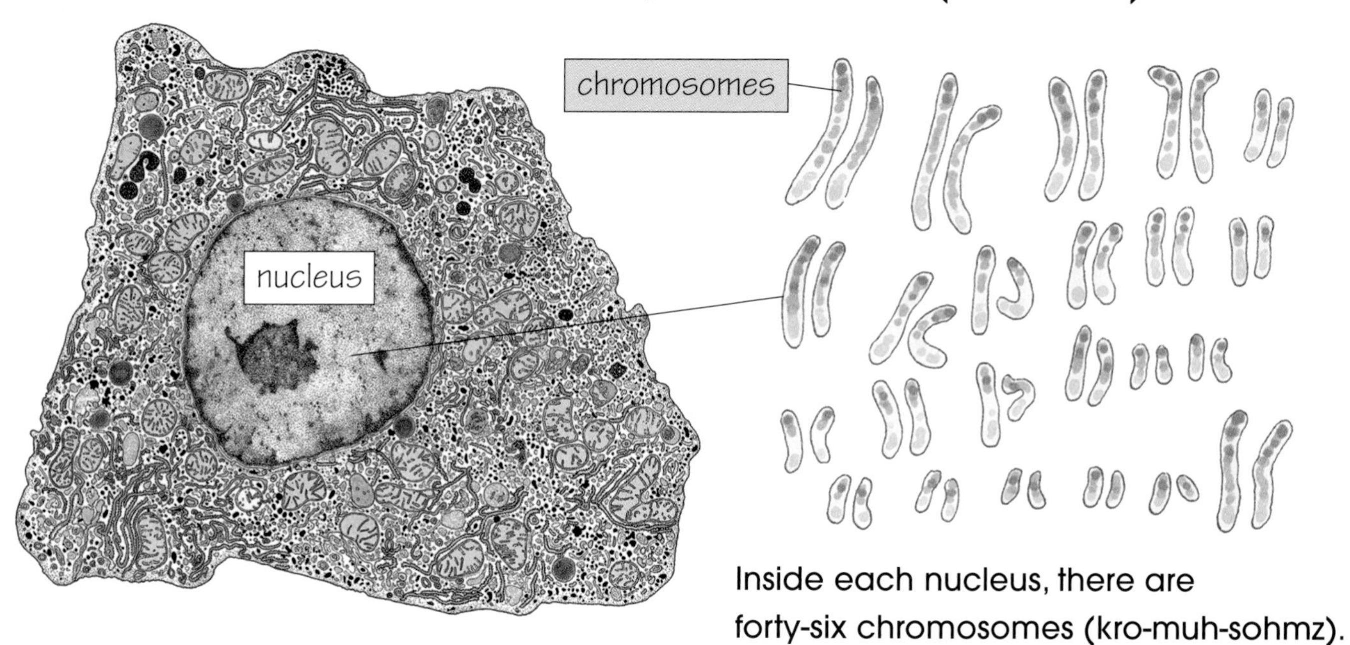

Inside each nucleus, there are forty-six chromosomes (kro-muh-sohmz).

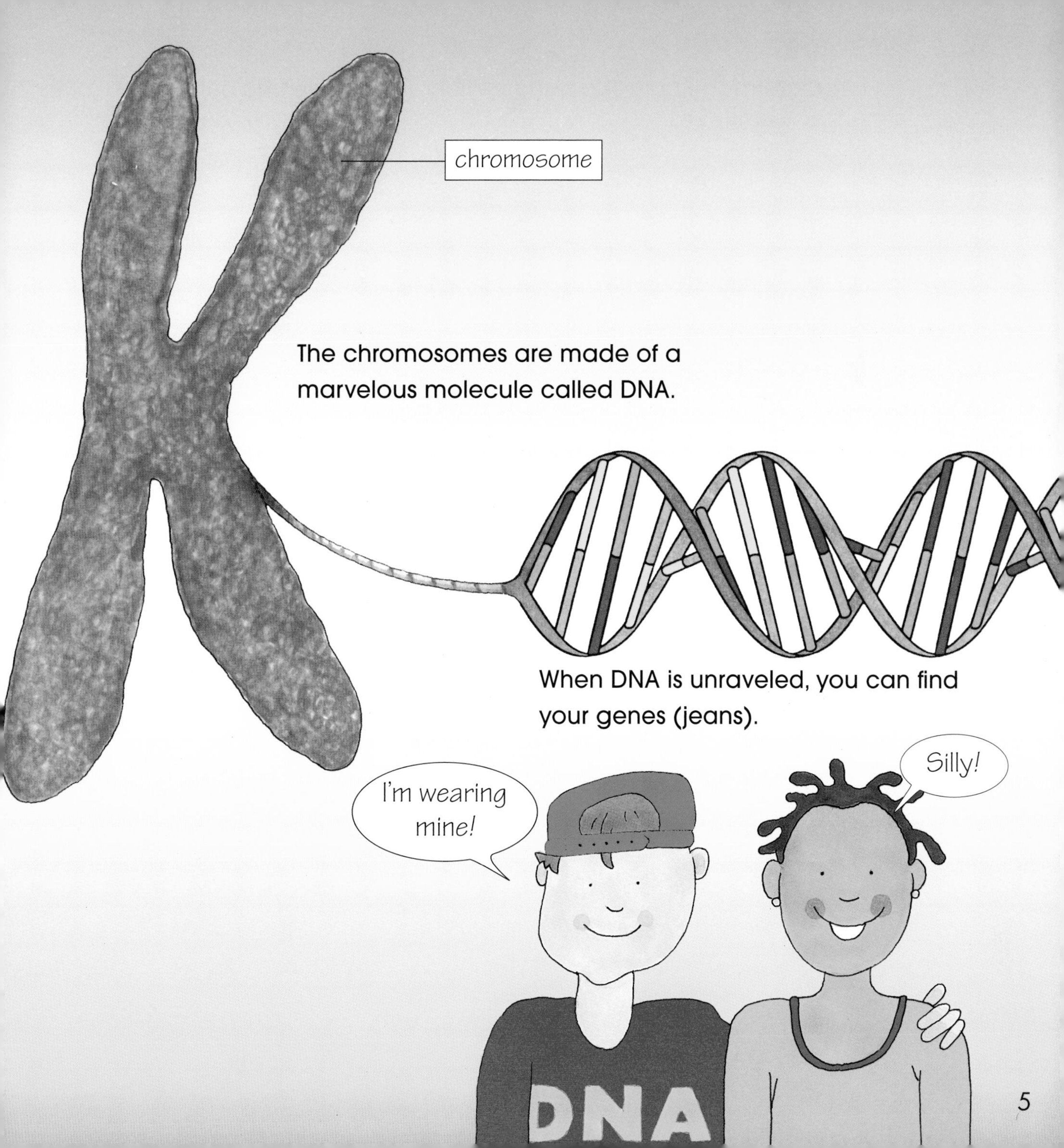

The chromosomes are made of a marvelous molecule called DNA.

When DNA is unraveled, you can find your genes (jeans).

Now you know where to find your genes. But what are they?

Genes are recipes for making proteins.

Proteins (pro-teens) are the most important and complicated chemicals in your body. They build your cells, and carry out the different jobs that your cells do.

Almost everything that is important in your cells is made up **of** proteins or is made **by** proteins!
lung cells

So, how do genes tell the cell to make proteins?

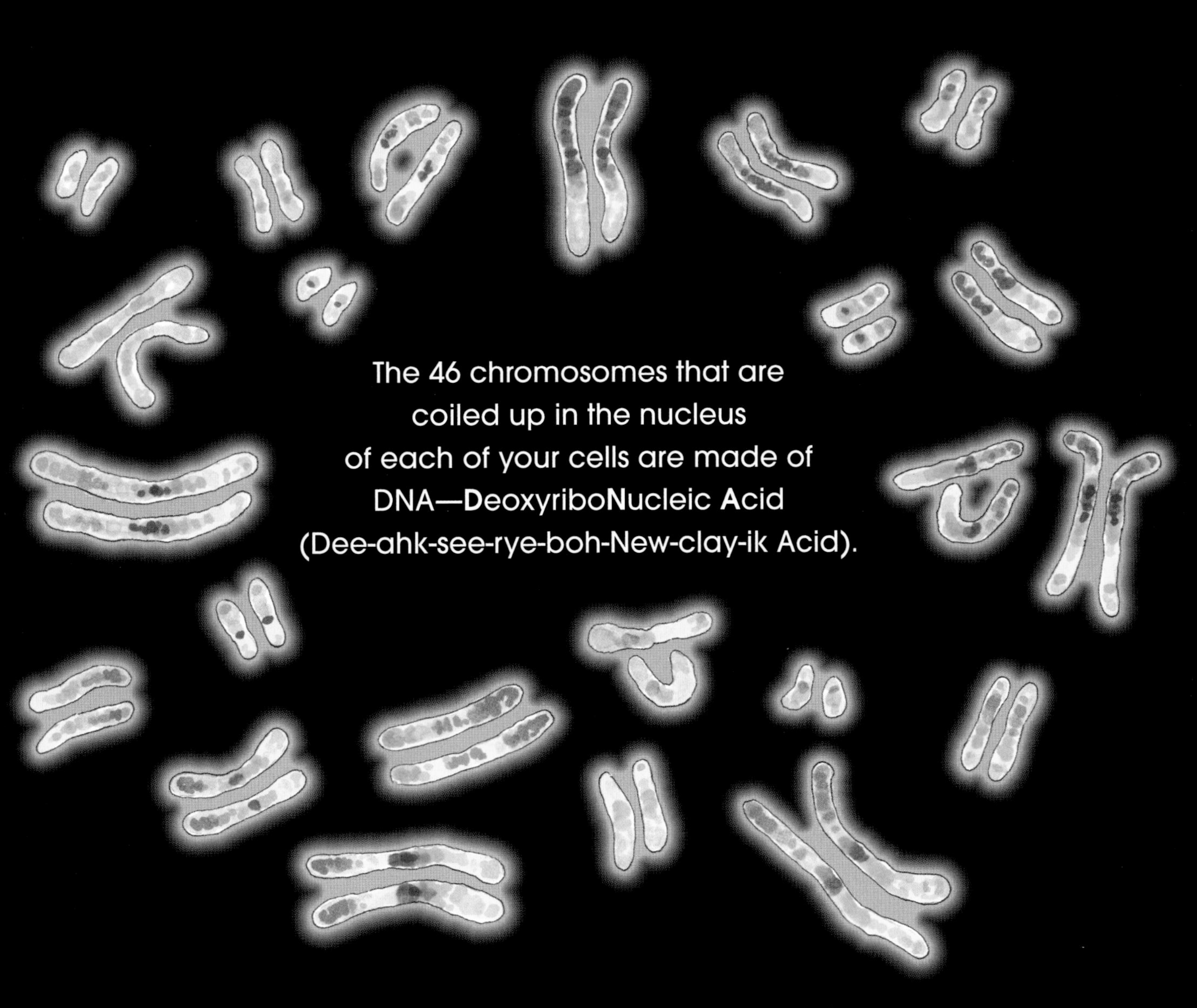

The 46 chromosomes that are coiled up in the nucleus of each of your cells are made of DNA—**D**eoxyribo**N**ucleic **A**cid (Dee-ahk-see-rye-boh-New-clay-ik Acid).

If genes are recipes for making proteins, DNA is the chemical language that spells them out.

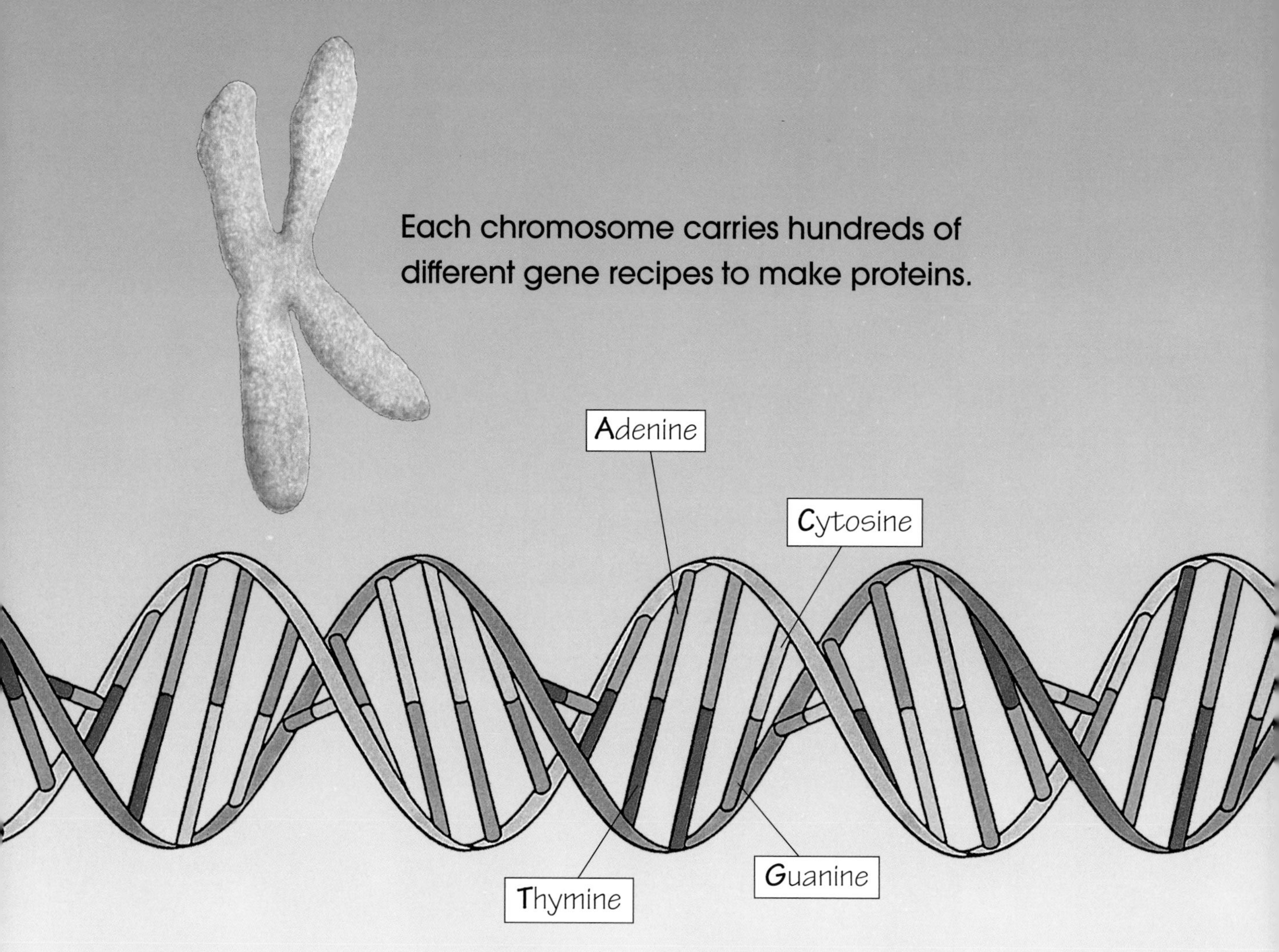

Each chromosome carries hundreds of different gene recipes to make proteins.
Adenine
Cytosine
Guanine
Thymine

The DNA language that makes the gene recipes uses just four different chemicals.

Adenine, **T**hymine, **C**ytosine, and **G**uanine—we will call them by their initial letters **A**, **T**, **C**, and **G** .

In this book, **A**denine is always shown as the color green, **T**hymine is red, **C**ytosine is yellow, and **G**uanine is blue.

If DNA is the language that writes your genes, you can think of **A**, **T**, **C**, and **G** as the letters of that language.

DNA is made of not one, but two strands. Each strand is coiled in a shape called a helix.

The two strands of DNA are joined together in a twisting turning spiral. This shape is called a **double** helix.

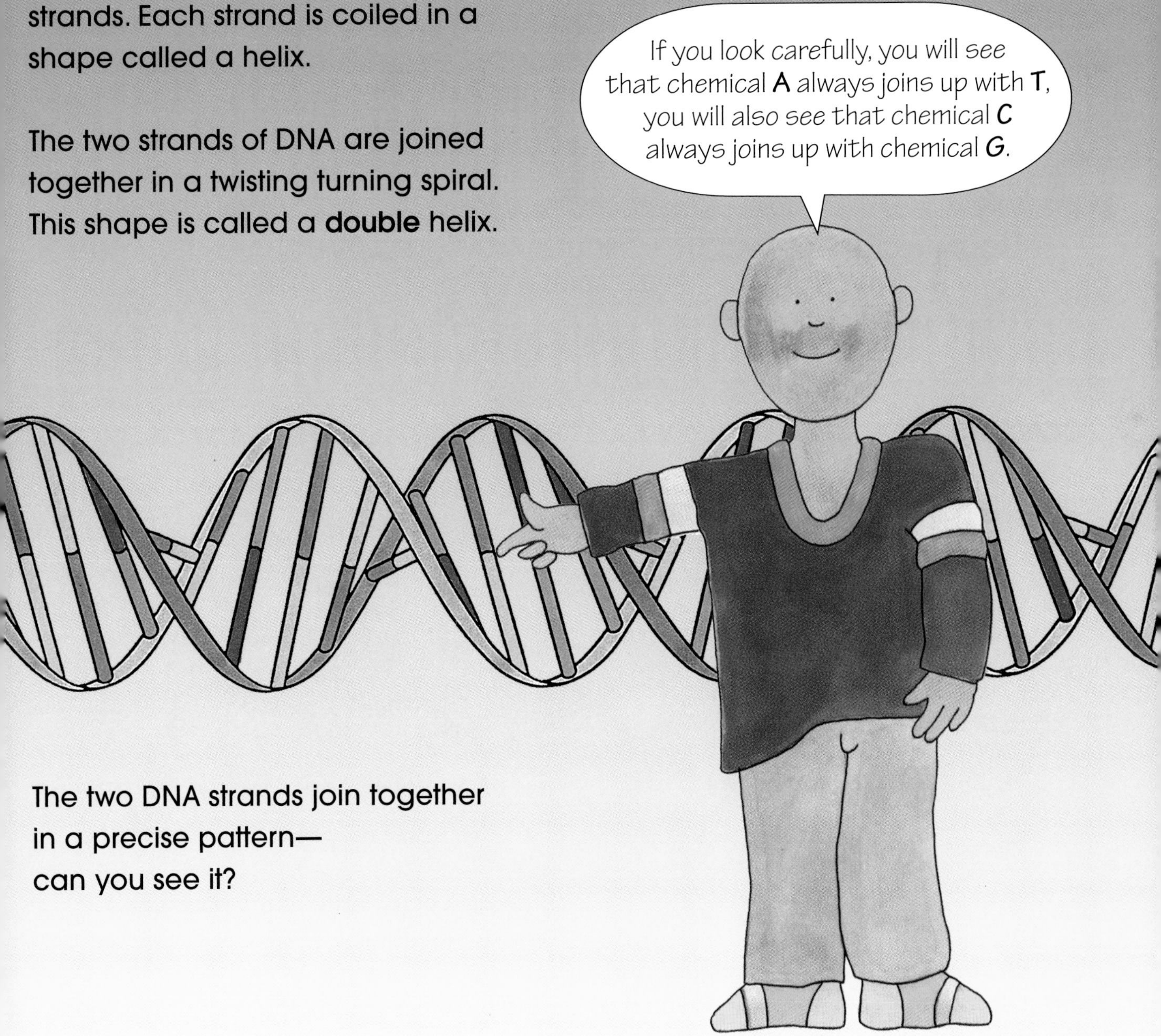

If you look carefully, you will see that chemical **A** always joins up with **T**, you will also see that chemical **C** always joins up with chemical **G**.

The two DNA strands join together in a precise pattern—
can you see it?

This is how a cell makes a protein.

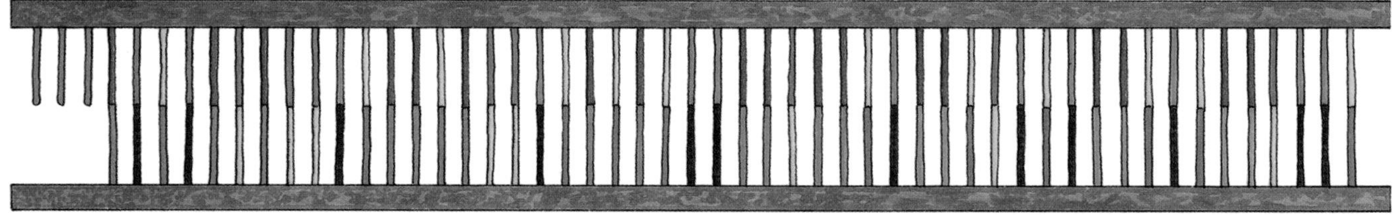

First, the small part of the DNA thread that contains the gene for that protein unwinds. Then a copy of that stretch of DNA is made.

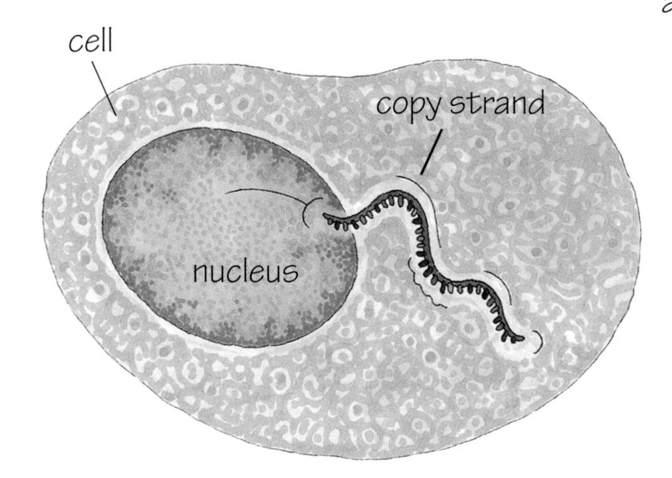

The DNA copy floats out of the nucleus and finds a ribosome (rye-boh-sohm), which is a place where proteins are made.

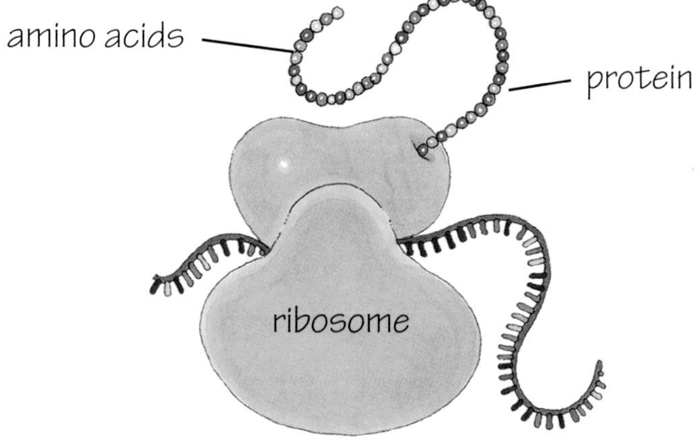

The DNA copy instructs the ribosome to make the protein.

The protein is assembled from amino acids that are found close to the ribosomes.

Look around at your friends and family. Think of the people that live in your neighborhood. Each and every one is different. And so are YOU. Why is that?

Well, everyone's mixture of genes is unique.

But how do you have a unique mixture of genes?

The first thing to know is that you have not one, but two sets of genes. One set is from your father and the other set is from your mother.

Each gene may be very slightly different from the other gene in the pair.

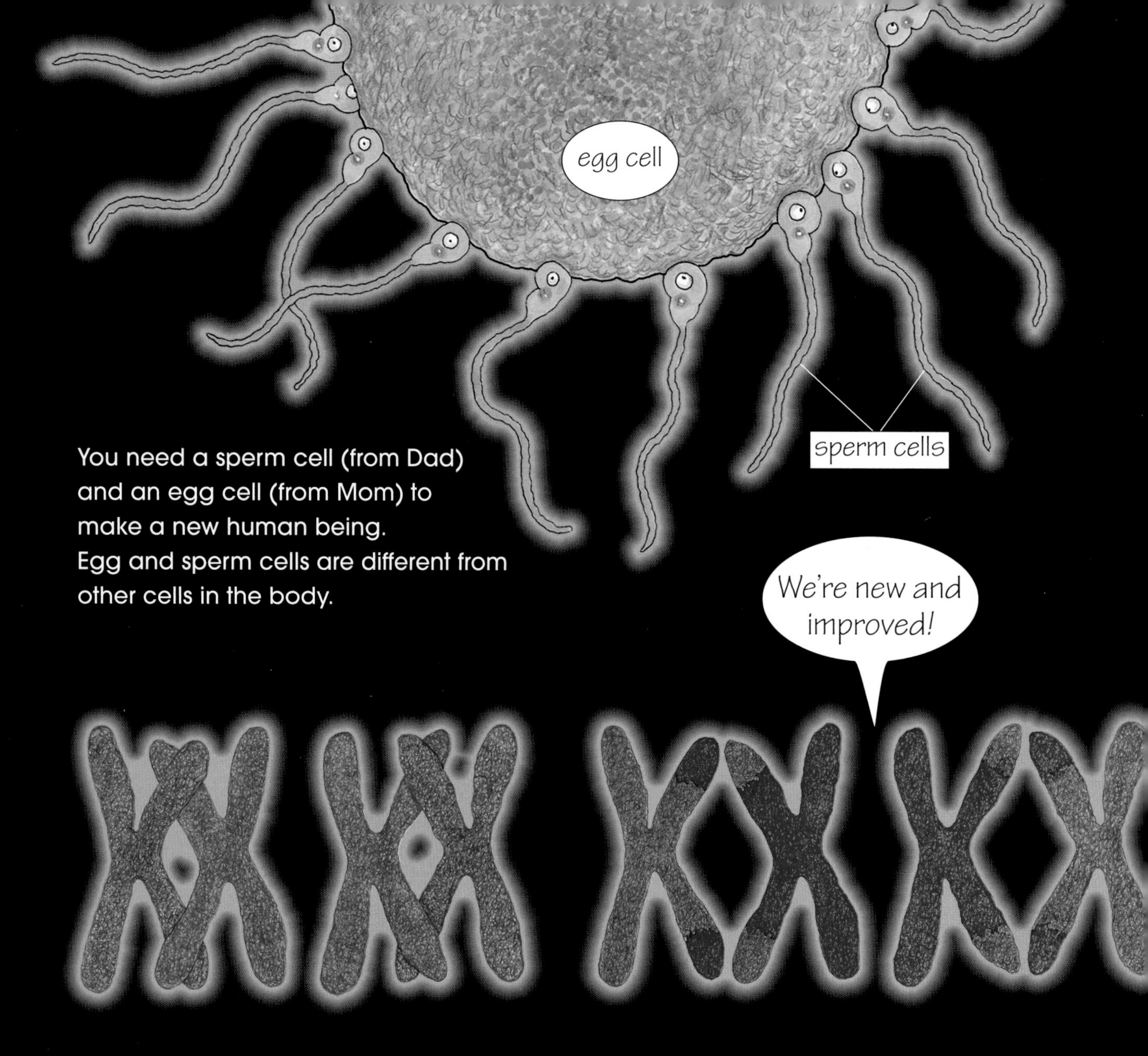

You need a sperm cell (from Dad) and an egg cell (from Mom) to make a new human being. Egg and sperm cells are different from other cells in the body.

Their chromosomes can wind around one another and do something rather strange. They swap some genes!

This means that every sperm cell or egg cell has a unique mixture of chromosomes and genes.

The cell begins to split in half. This will make two cells with identical DNA instructions.

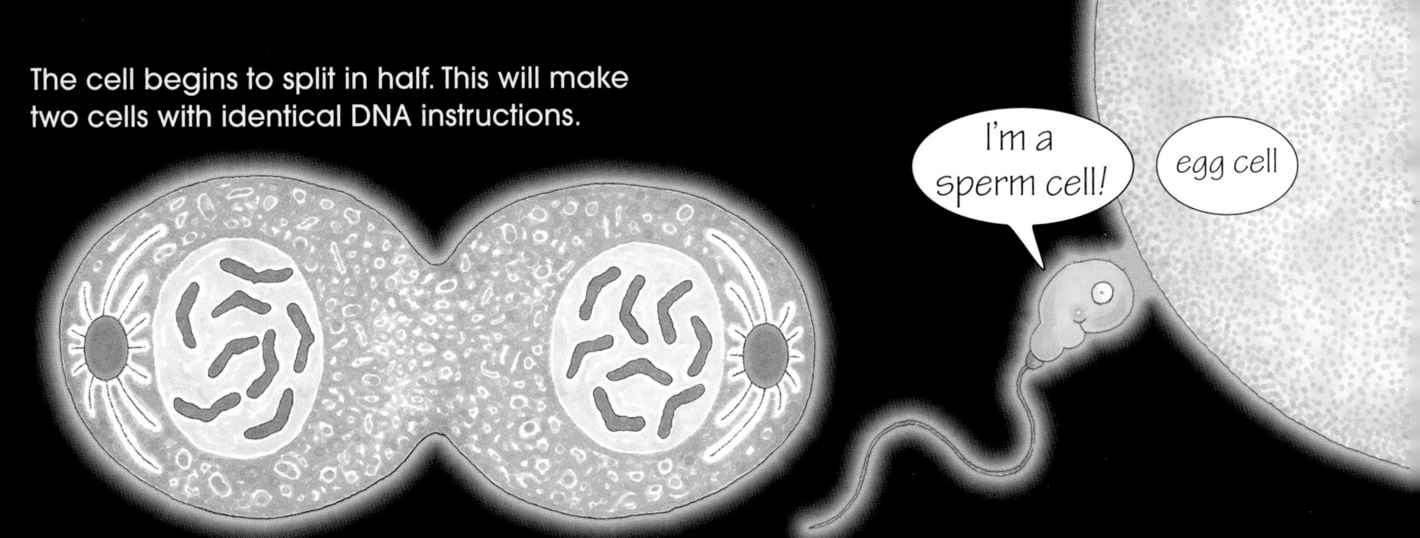

And by the time they are ready to make a new human being, sperm cells and egg cells have only 23 chromosomes each, instead of 46. You were made when one egg cell with 23 chromosomes joined with one sperm cell also with 23 chromosomes.

So the first cell that was you had 46 chromosomes—23 from the sperm and 23 from the egg.

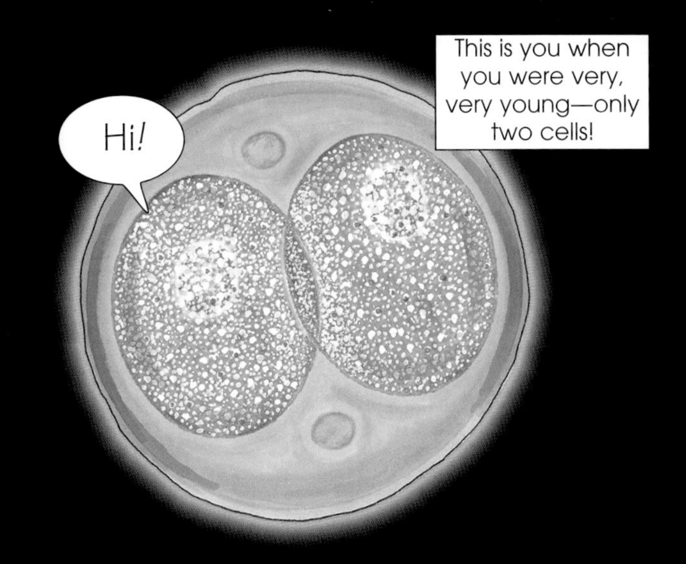

You became one brand new cell —a cell that was uniquely **you.**

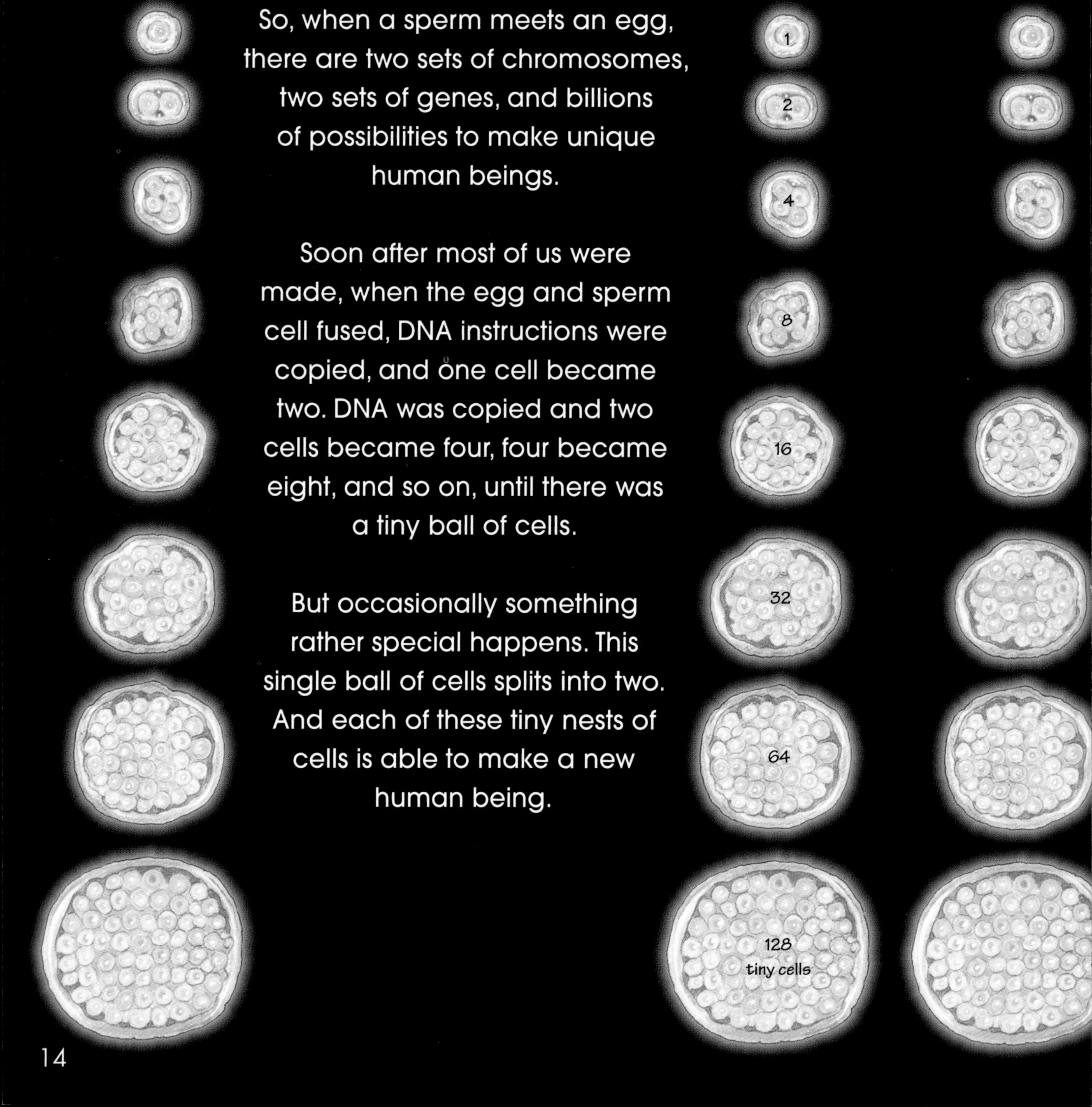

So, when a sperm meets an egg, there are two sets of chromosomes, two sets of genes, and billions of possibilities to make unique human beings.

Soon after most of us were made, when the egg and sperm cell fused, DNA instructions were copied, and one cell became two. DNA was copied and two cells became four, four became eight, and so on, until there was a tiny ball of cells.

But occasionally something rather special happens. This single ball of cells splits into two. And each of these tiny nests of cells is able to make a new human being.

Then we have identical twins with identical genes!

Their identical genes mean that twins look much the same, even when they are very old. Identical twins have identical height, fingerprints, and hair color. Their brainwaves, IQ, and weight are also similar.

Not all twins have identical genes, though. Sometimes the mother makes two eggs instead of one. If two sperm cells fuse with those two eggs, the twins will be like any other brother or sister, except that they share a birth date.

identical twins

non-identical twins

The effect of some genes is easy to recognize.

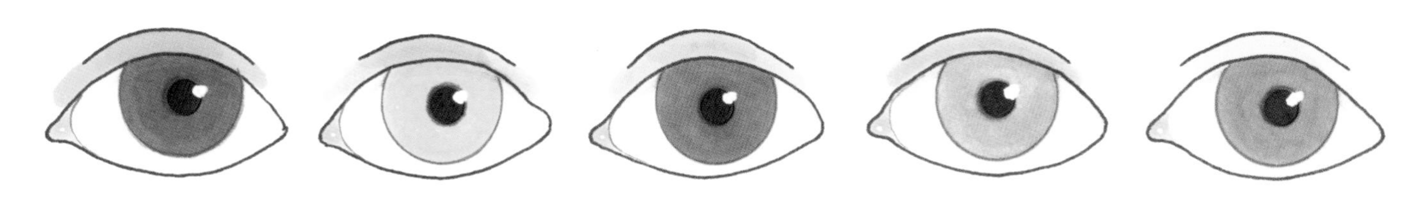

Genes make human eyes black, blue, brown, green, hazel, and many shades between.

Human hair can be curly, fine, or frizzled; blond, brown, black, or even red, thanks to your genes.

Human skin can be many different shades of brown, reddish, pinkish, olive, or pale yellow.

In fact, tens, hundreds, and even thousands of genes influence most of the things that make us all different from each other.

But genes are only **one** reason for the diversity of human beings. Just as important is the world around you.

The way you live, the food you eat, your school, your family and friends, good and bad things that happen to you all work alongside your genes to make you as you are today—and will be tomorrow—even if you are an identical twin!

What do you think is the most important difference between boys and girls?

It is all in a gene!

Chromosomes from a girl can be sorted into 23 pairs of similar size. But in the cells of any boy, you find 22 pairs of identical-looking chromosomes and one pair that look quite different from each other.

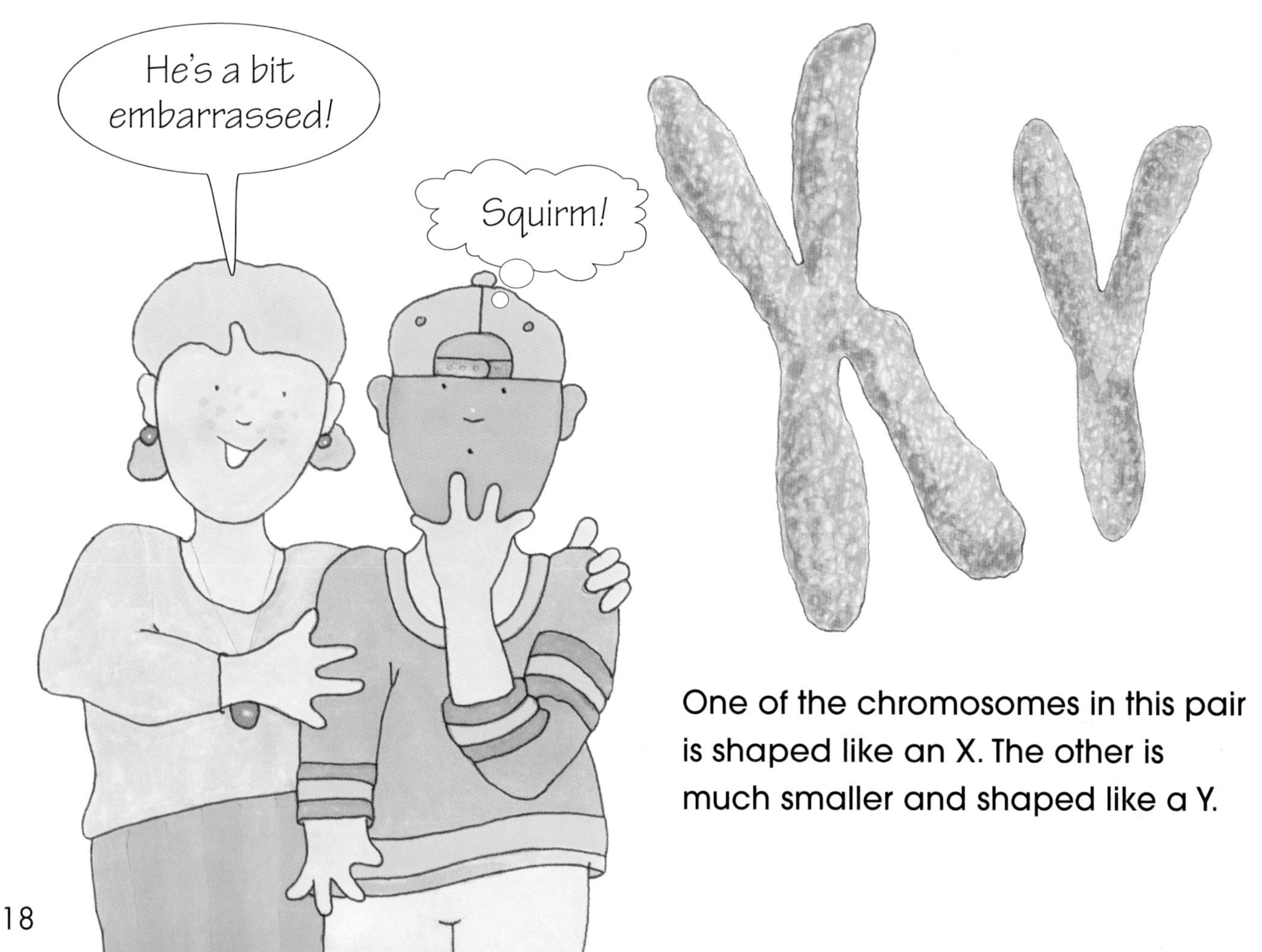
He's a bit embarrassed!
Squirm!

One of the chromosomes in this pair is shaped like an X. The other is much smaller and shaped like a Y.

At the beginning, humans all develop in the same way. But something changes after a few days. If the embryo has two X chromosomes, it will develop into a girl baby. If the embryo has one X and one Y chromosome, it will begin to become a boy.

This is because one of the genes on that Y chromosome is rather special.

This gene sends a chemical signal to a little group of cells in the tiny embryo. These cells begin to turn into sperm cells.

Many more genes on other chromosomes make many more proteins that make boys look the way they do. However, without that gene on chromosome Y...

XX
I'm a girl!
XY
I'm a boy!
...we would all be **girls**!
XX
OR
XY

One makes a girl, one makes a boy!

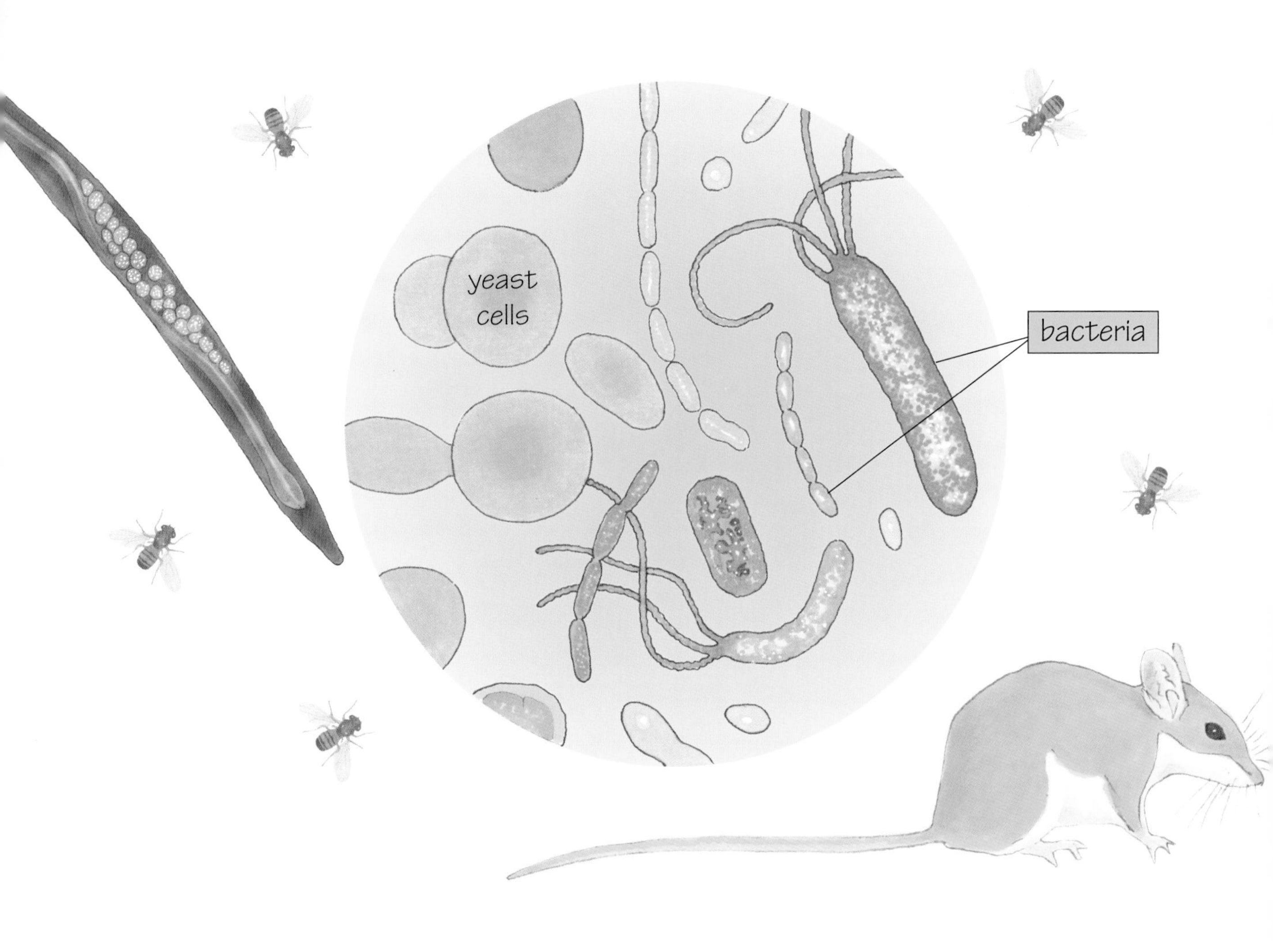

You might think that humans are very different from mice and flies and worms. And of course you must think that you are enormously different from single-cell creatures such as yeast and bacteria.

But by studying genes, scientists have learned that flies, worms, mice, and even microscopic yeast are quite closely related to you! You have many genes that are almost exactly the same as their genes.

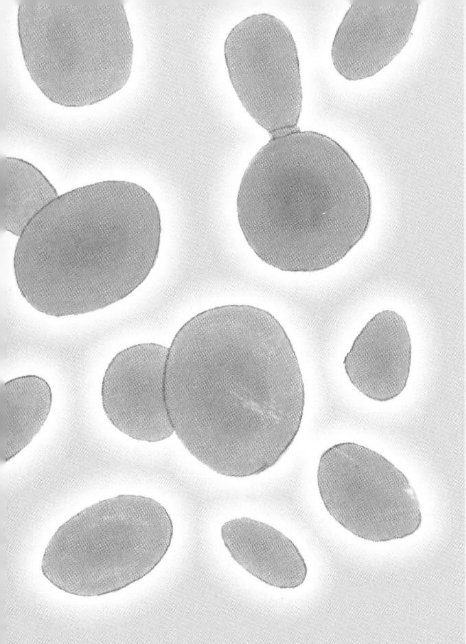

Each simple yeast cell has about 6,000 genes.

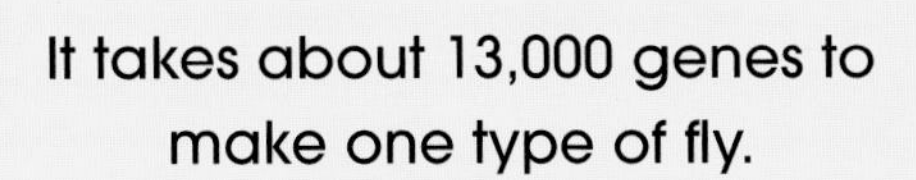

It takes about 13,000 genes to make one type of fly.

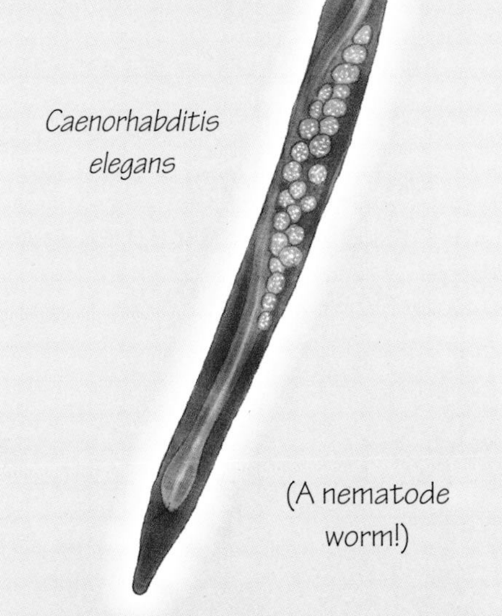

About 18,000 genes make one type of worm.

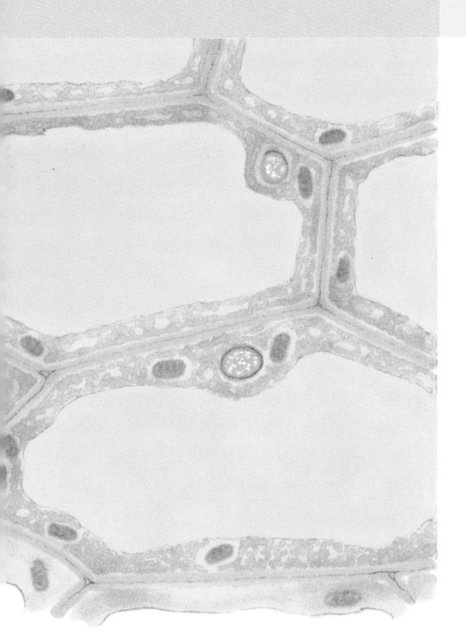

About 26,000 genes are found in more complicated plant cells.

So how many genes do you think it takes to make you?

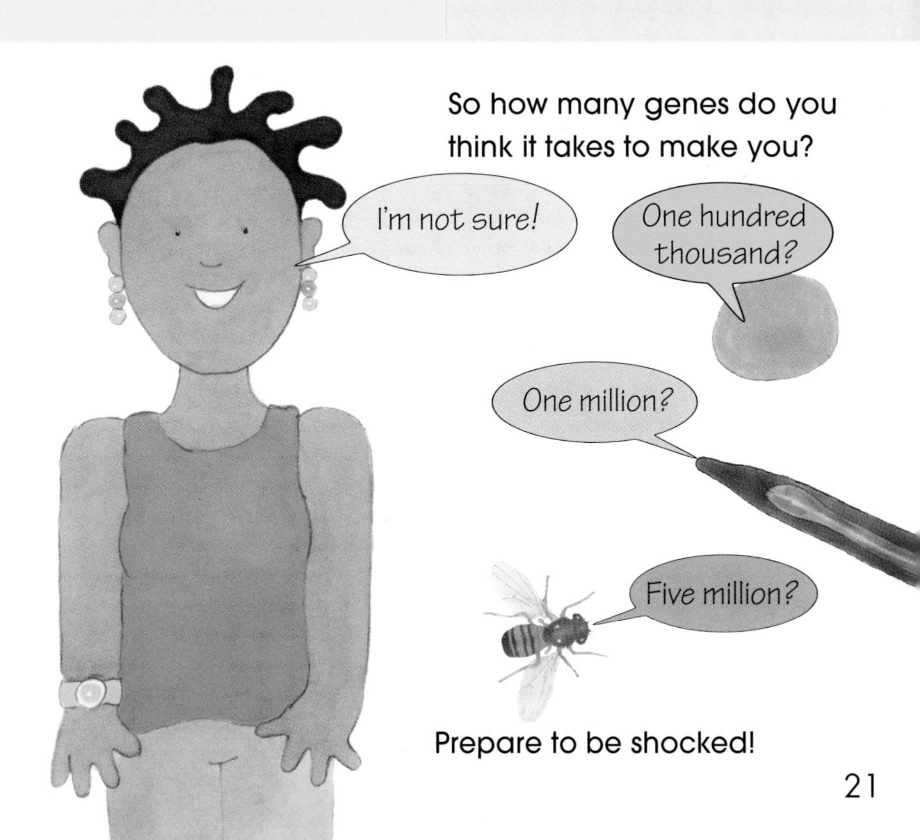

I'm not sure!
One hundred thousand?
One million?
Five million?

Prepare to be shocked!

About 30,000 genes are all that is needed to make a complex human being like you! And you probably have only about a few hundred genes more than a mouse.

Look at how the insides of a mouse are organized. See how similar it is to yours. This is because you have so many genes in common.

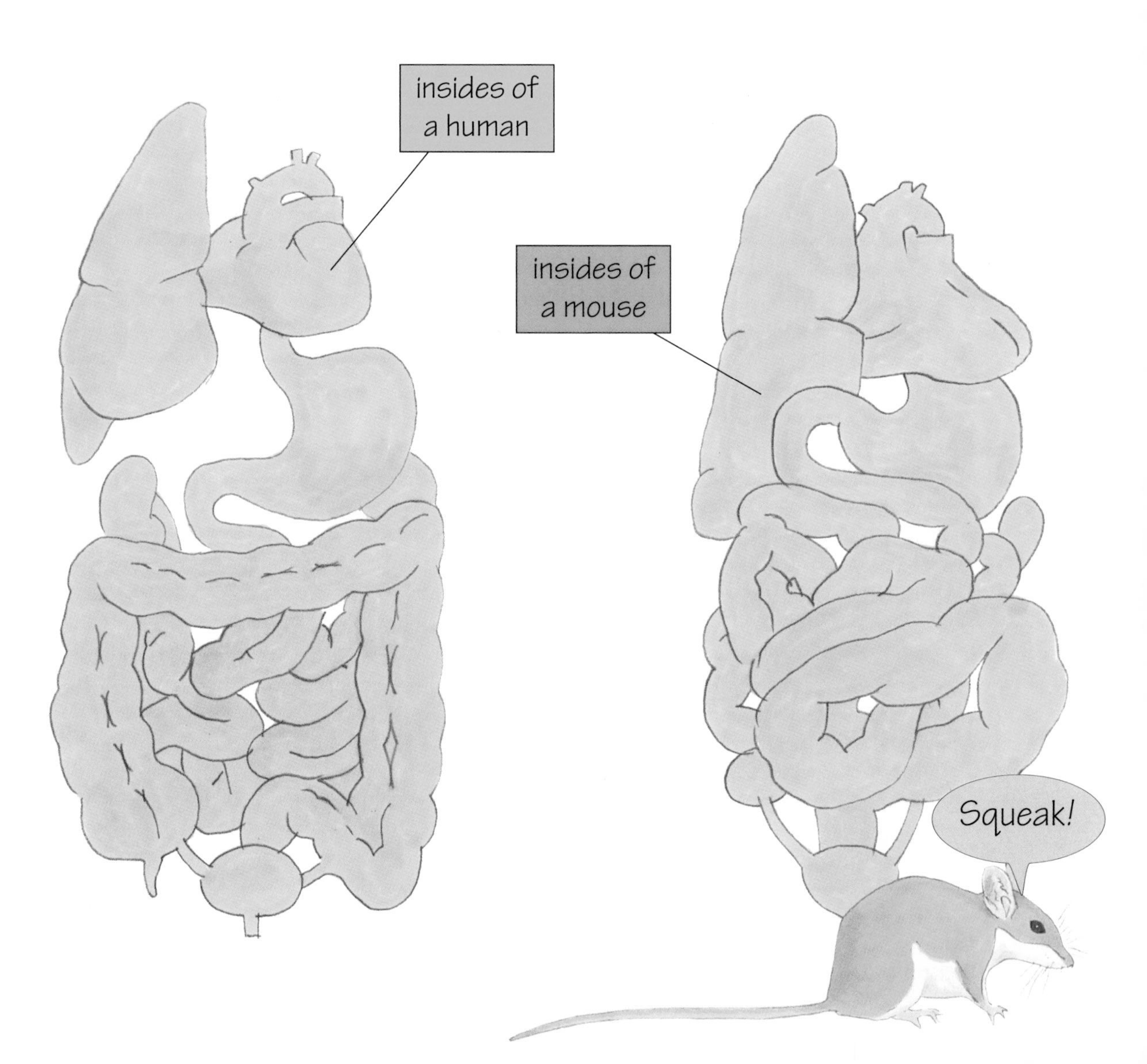

When scientists looked carefully at the genes that we humans possess, they had another surprise.

Over 200 of your genes are the same as those found in bacteria. But flies, worms, and plants do not have these genes.

Somehow, when human beings first began to live on Earth, some DNA from bacteria cells became mixed in with the human DNA. Maybe it came from the friendly bacteria that live in our intestines. This DNA did no harm, it might have even been useful. So there it has stayed.

Because each cell has a complete set of genes, it is possible to make a whole creature from just one single cell.

This is called a clone. Scientists have been able to clone animals such as mice and sheep and cows. One way scientists have made an animal clone is called nuclear transfer (new-clee-are tranz-fur).

They use the cell nucleus of an animal that has already been born. This nucleus is placed in an empty egg cell and allowed to develop in a female animal of the same species.

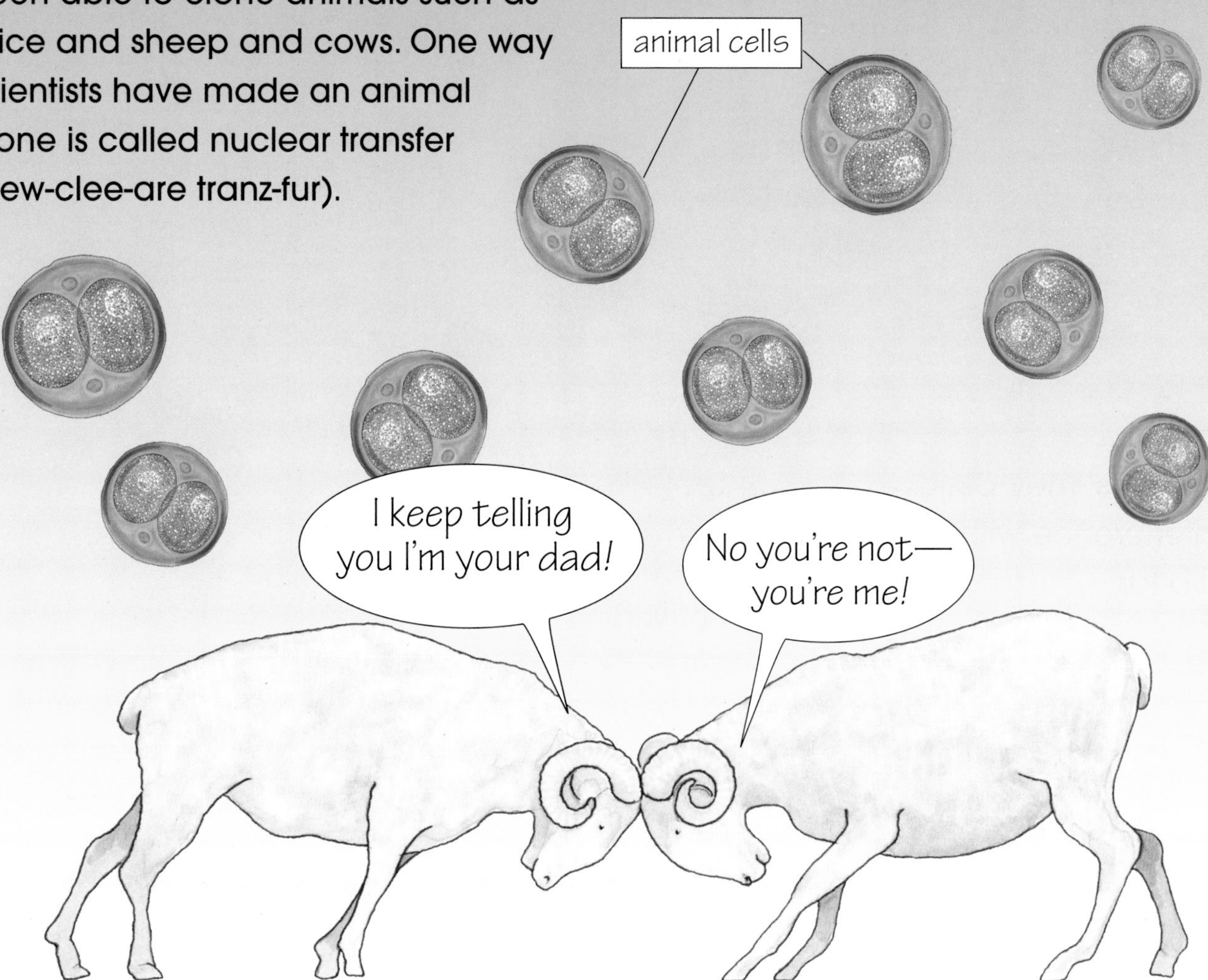

Cloning is not an easy thing to do. Many of the cloned creatures die when they are just a few cells old, some die before they are born, and others die soon after. At the moment, only one in every 100 cloned embryos will survive and live.

And many people are worried about the prospect of cloning animals. They believe that human interference with Nature has gone too far.

Others believe that cloning may be an important way to save some endangered species, or even recreate extinct species.

Maybe you think your very own clone would be a great idea? Let's see what would happen during nuclear transfer.

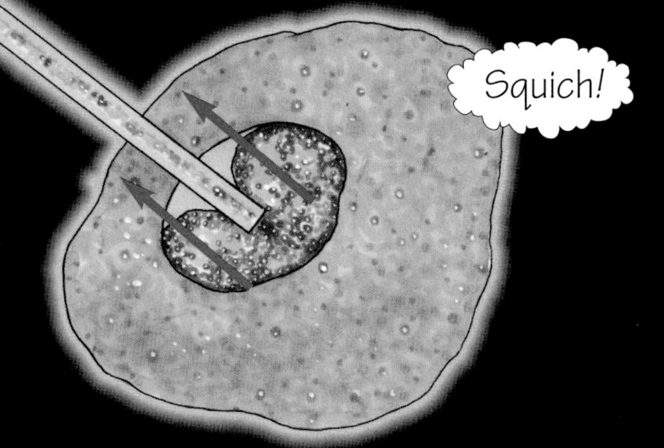

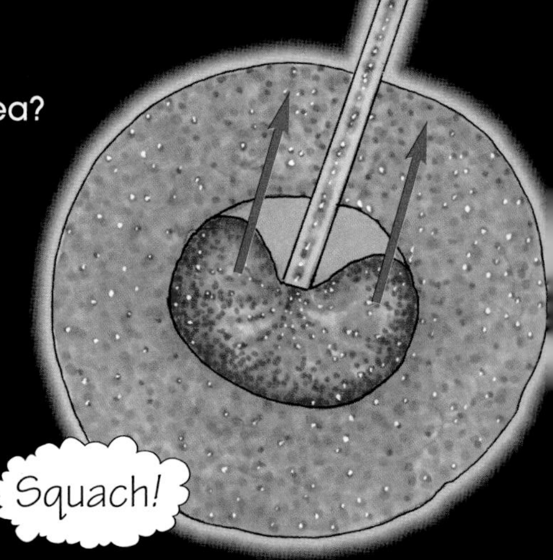

Just imagine—take one of your cells and remove its nucleus. This single nucleus contains all the instructions to make you (again!). Remove the nucleus from a human egg cell (from someone else or you!).

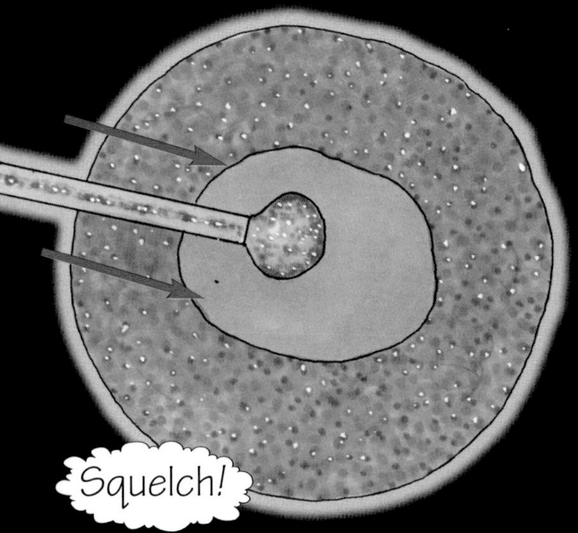

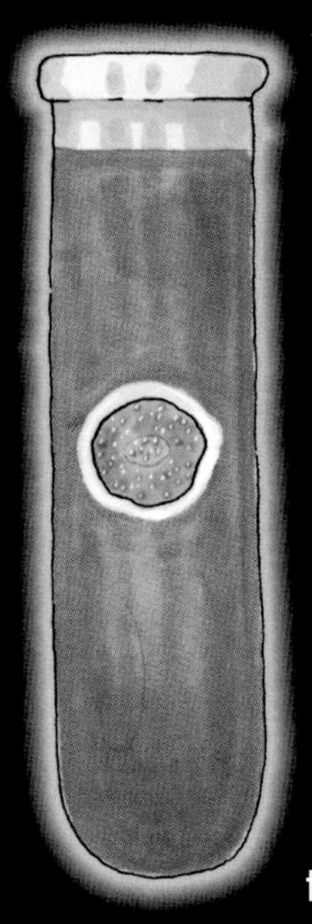

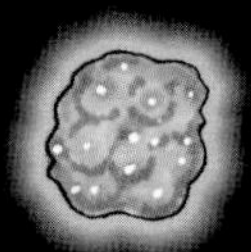

Insert the DNA from your cell nucleus into the egg cell. Put this cell into a test tube.

Add special chemicals to the egg cell. Your DNA is copied. The egg cell starts to grow and becomes two cells, then four, then eight, then 16, and so on.

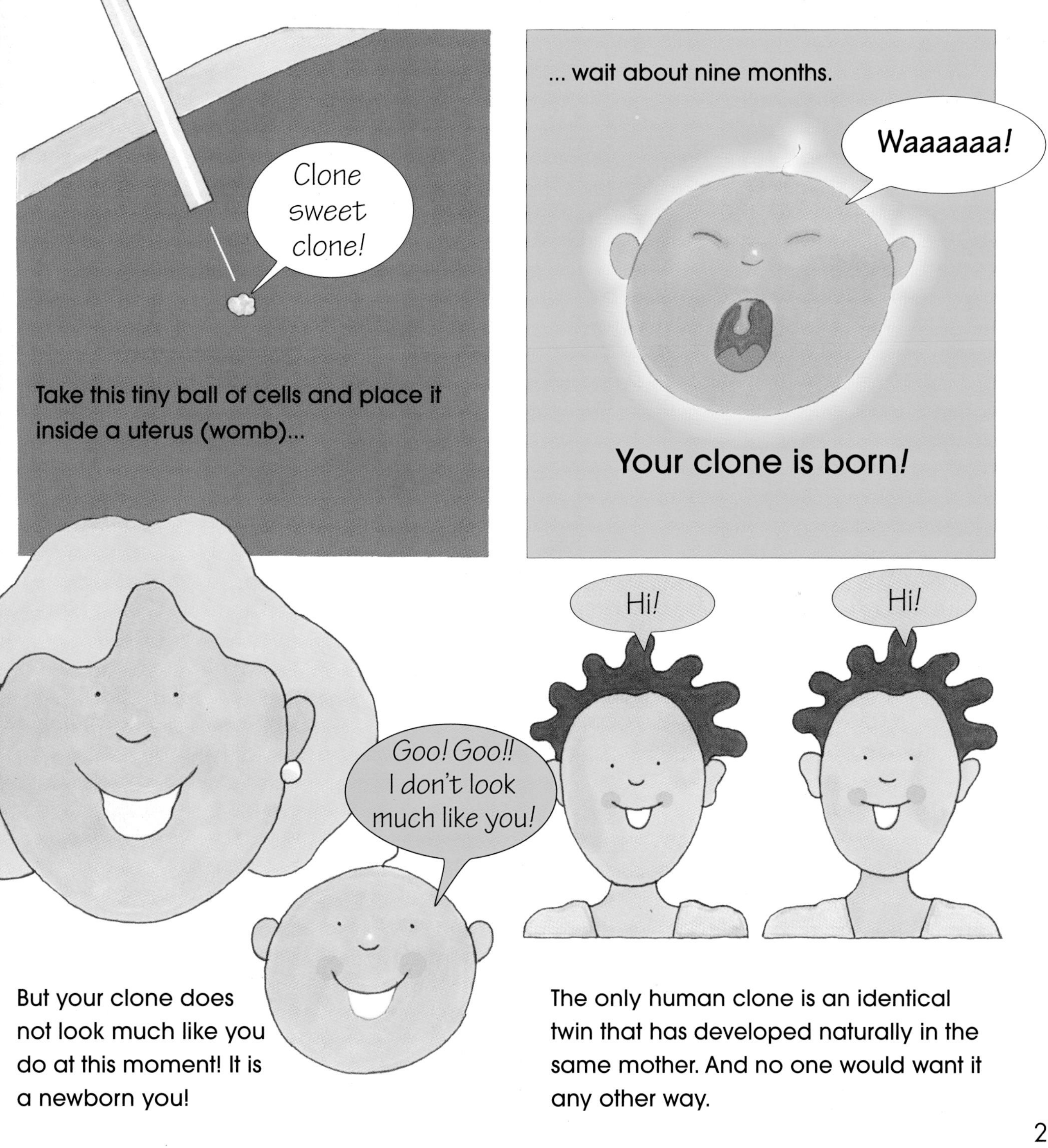
Clone sweet clone!
Take this tiny ball of cells and place it inside a uterus (womb)...
... wait about nine months.
Waaaaaa!
Your clone is born!
Goo! Goo!! I don't look much like you!
But your clone does not look much like you do at this moment! It is a newborn you!
Hi!
Hi!
The only human clone is an identical twin that has developed naturally in the same mother. And no one would want it any other way.

Hundreds of diseases are caused by faulty genes that make faulty proteins (or cannot make their protein at all).

You may have heard about some of them. Sickle cell anemia and cystic fibrosis are two genetic diseases that make children ill.

Studying our genes helps us understand these illnesses and find new treatments and cures.

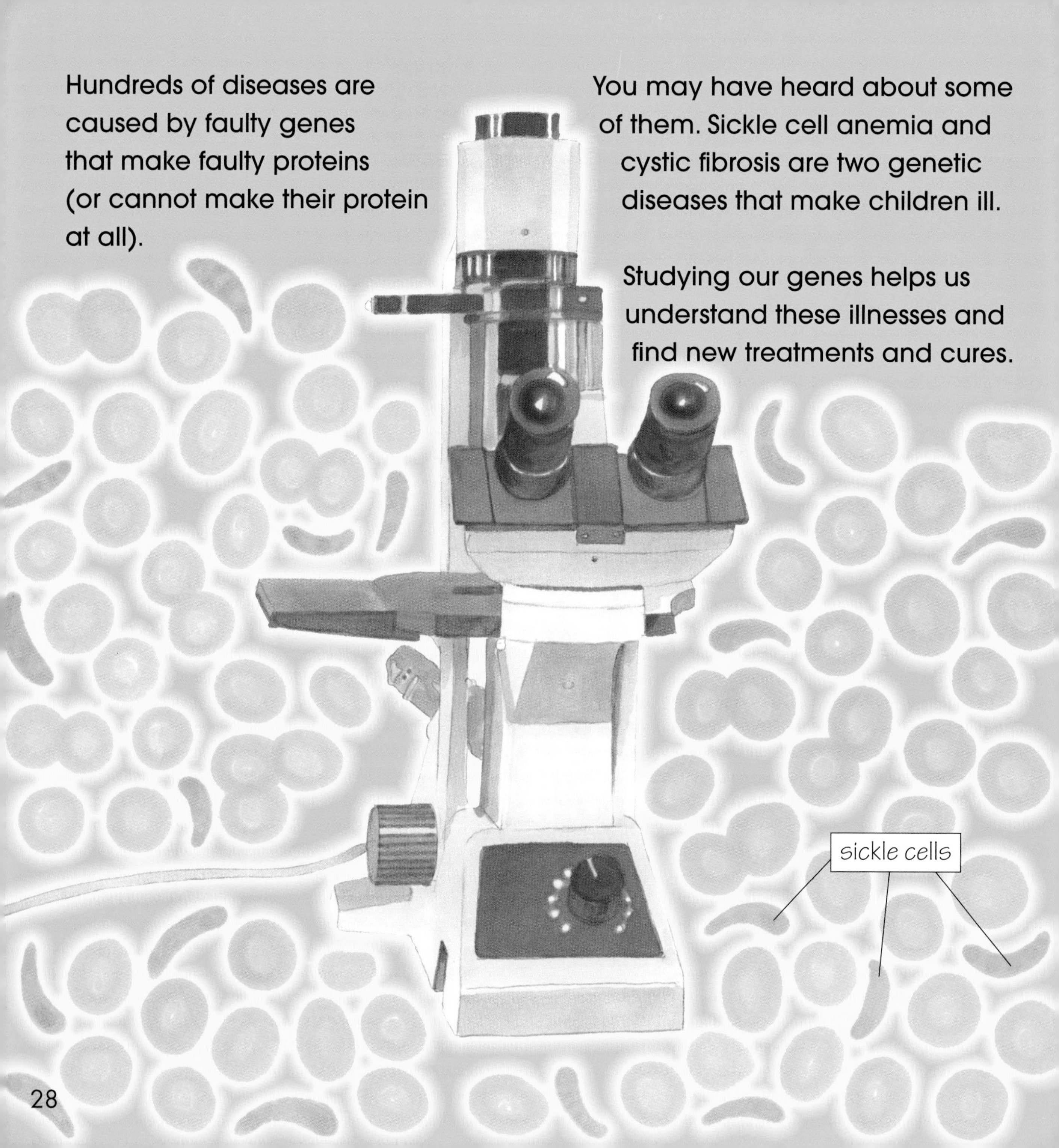

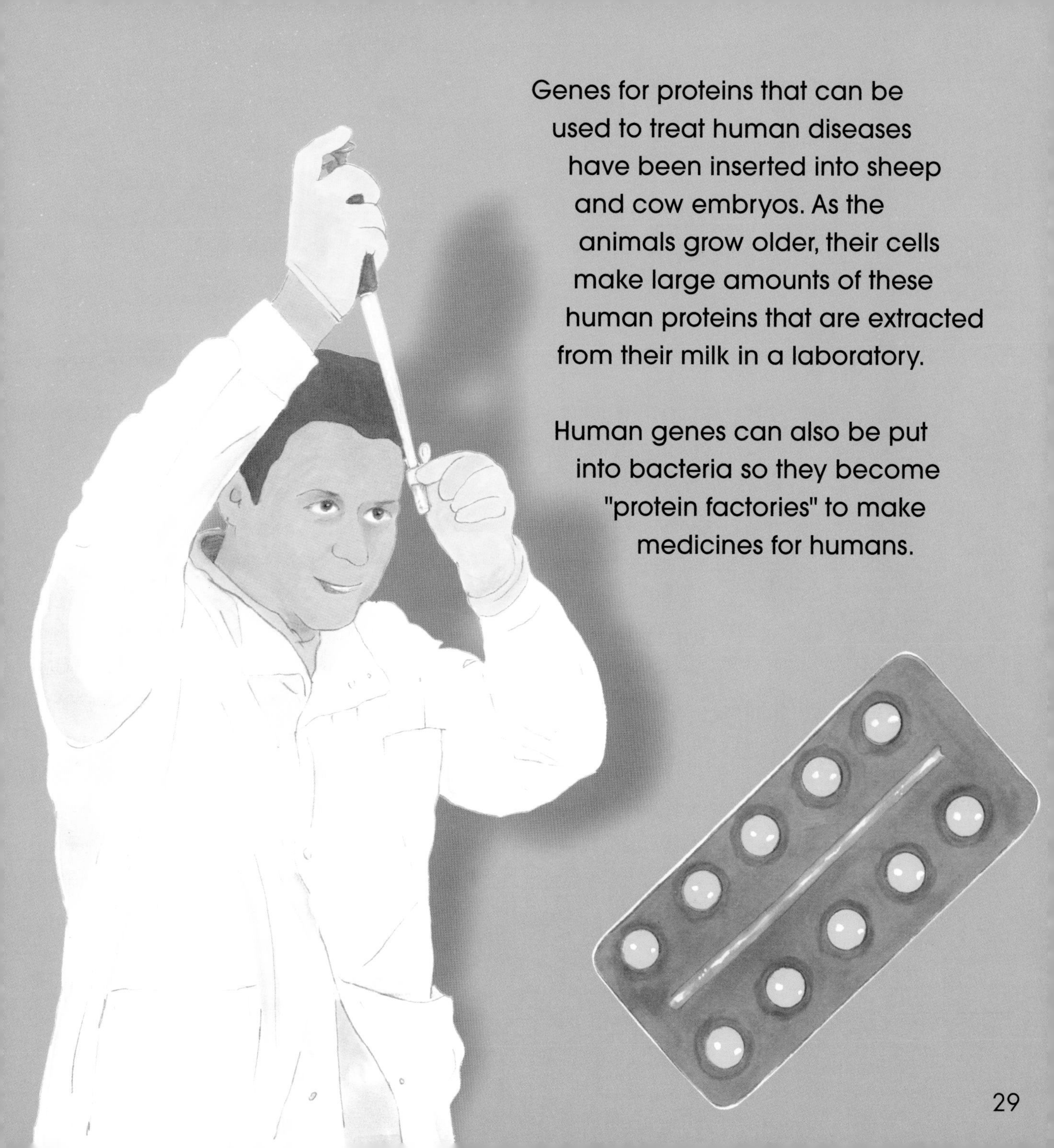

Genes for proteins that can be used to treat human diseases have been inserted into sheep and cow embryos. As the animals grow older, their cells make large amounts of these human proteins that are extracted from their milk in a laboratory.

Human genes can also be put into bacteria so they become "protein factories" to make medicines for humans.

We have told you that you share your genes with every other life form on Earth. Your genes tell the story of how life evolved from a single-cell creature some four thousand million years ago. Some of those genes that were in that first creature are the same as the genes that make proteins in your cells today.

You share almost all your genes with all the animals that have backbones. But why are you so different from mice, cats, dogs, and chimpanzees when most of your genes are the same?

The answer to that question will unlock one of Nature's greatest secrets.

We hope you enjoyed your cells!